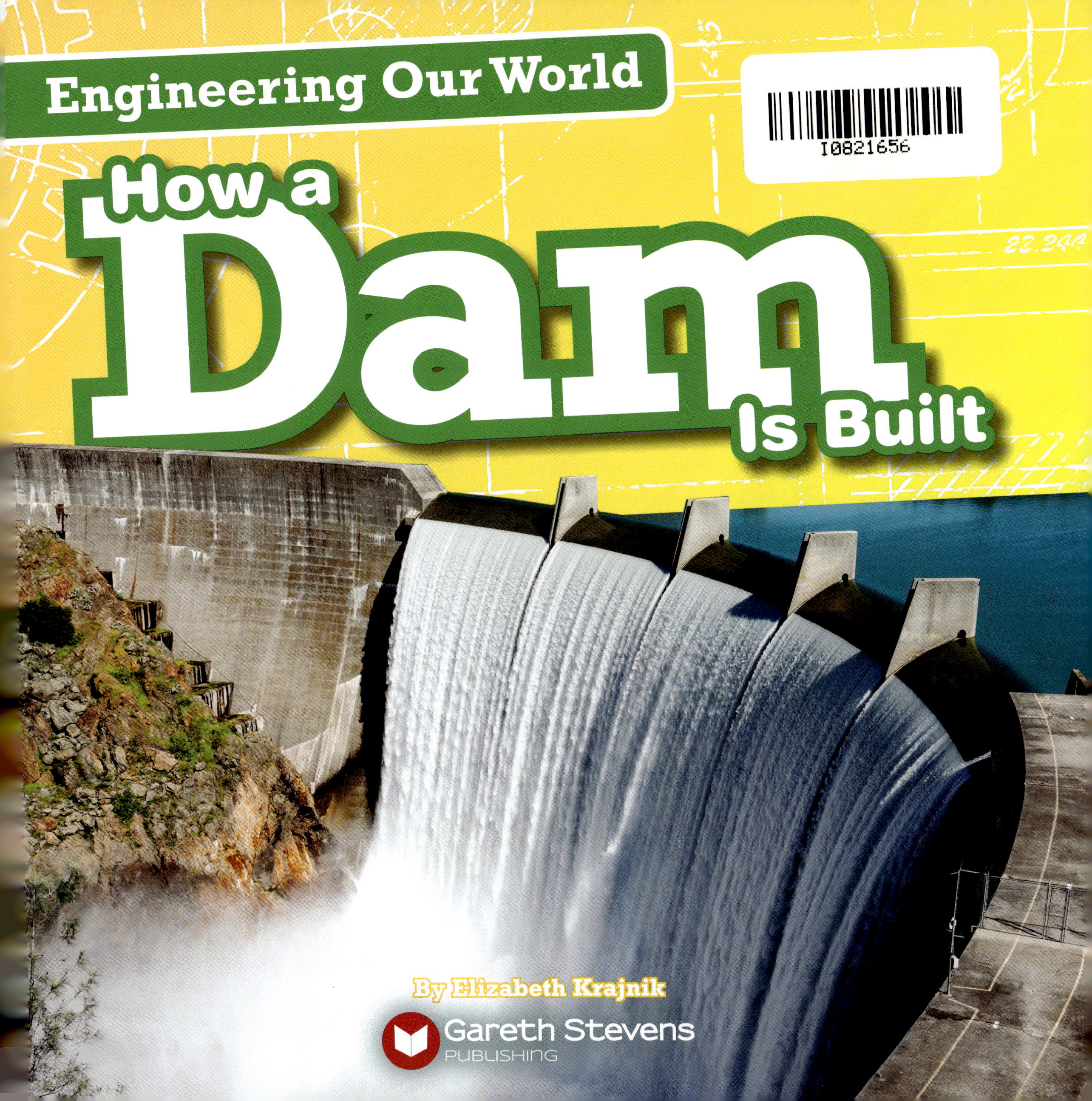
Engineering Our World
How a Dam Is Built
By Elizabeth Krajnik
Gareth Stevens
PUBLISHING
I0821656

Please visit our website, www.garethstevens.com. For a free color catalog of all our high-quality books, call toll free 1-800-542-2595 or fax 1-877-542-2596.

Cataloging-in-Publication Data

Names: Krajnik, Elizabeth.
Title: How a dam is built / Elizabeth Krajnik.
Description: New York : Gareth Stevens Publishing, 2021. | Series: Engineering our world | Includes glossary and index.
Identifiers: ISBN 9781538246986 (pbk.) | ISBN 9781538247006 (library bound) | ISBN 9781538246993 (6 pack)
Subjects: LCSH: Dams–Juvenile literature. | Dams–Design and construction–Juvenile literature.
Classification: LCC TC541.K73 2021 | DDC 627'.8–dc23

First Edition

Published in 2021 by
Gareth Stevens Publishing
111 East 14th Street, Suite 349
New York, NY 10003

Designer: Sarah Liddell
Editor: Monika Davies

Photo credits: Cover, p. 1 Gary Saxe/Shutterstock.com; background Jason Winter/Shutterstock.com; p. 5 DEA/G. NIMATAALLAH/Contributor/De Agostini/Getty Images; p. 7 (arch dam) JJFarq/Shutterstock.com; p. 7 (embankment) M-ken/Shutterstock.com; p. 7 (buttress dam) ODD ANDERSEN/Staff/AFP/Getty Images; p. 7 (gravity dam) Sveta Imnadze/Shutterstock.com; p. 9 DeStefano/Shutterstock.com; p. 11 The Denver Post/Contributor/Denver Post/Getty Images; p. 13 Zyphyrus/Shutterstock.com; p. 15 Chaikom/Shutterstock.com; p. 17 nootprapa/Shutterstock.com; p. 19 Agence France Presse/Contributor/AFP/Getty Images; p. 20 (gravel) bogdan ionescu/Shutterstock.com; pp. 20, 21 (wooden sticks) Thanwa photo/Shutterstock.com; p. 20 (sand) images and videos/Shutterstock.com; p. 20 (bucket) Ilya Andriyanov/Shutterstock.com; p. 21 (plastic container) HeinzTeh/Shutterstock.com.

Printed in the United States of America

Contents

Words in the glossary appear in **bold** type the first time they are used in the text.

Holding Back

A dam is a structure, or something built, that's used to retain, or hold back and store, water. Dams can be constructed across streams and rivers. People often build dams to get drinking water and water for **irrigation**. Dams can also help create **hydroelectric** power.

Many people are involved in building a dam. **Engineers** and city planners first come up with the building plan. They may talk to scientists to make sure building the dam won't hurt the **environment**. Then, construction workers begin the work to build the dam.

CORNALVO DAM IS AN ANCIENT ROMAN DAM IN SPAIN. IT'S ONE OF THE OLDEST MAN-MADE DAMS STILL IN USE TODAY!

Types of Dams

Before building a dam can begin, engineers have to decide what type of dam to build. Many modern dams are built partly or entirely out of concrete. There are three common types of concrete dams: gravity, arch, and buttress.

A gravity dam usually looks like a straight, strong wall of concrete across a valley. An arch dam is a curved, almost U-shaped structure of concrete to hold back water. A buttress dam is supported, or held up, by commonly triangular-shaped blocks made of concrete.

Building Blocks

Concrete is a mix of water, stones, sand, and a soft gray powder that becomes very hard and strong when it dries.

EMBANKMENT DAMS ARE ANOTHER COMMON TYPE OF DAM. THESE DAMS ARE MADE OF NATURAL MATERIALS, SUCH AS ROCK AND SOIL.

EMBANKMENT DAM

GRAVITY DAM

BUTTRESS DAM

ARCH DAM

Parts of a Dam

There are many types of dams, but all of them have common parts. On the upstream side of a dam, the bottom part is called the heel. This is where the river first touches the dam. The toe of a dam is on the downstream side.

Inside the dam is a spillway. The spillway is where extra water from the dam's **reservoir** spills out. Some dams also have a crest, which is the top, flat part of the dam. People can walk or drive on the dam's crest.

Building Blocks

Sand, soil, and mud can build up at the bottom of a dam's reservoir. A sluiceway or sluice gate can help clear out these pieces of matter from the reservoir.

DAMS SOMETIMES HAVE A GALLERY, OR TUNNEL, TO ALLOW WORKERS TO CHECK FOR CRACKS IN THE CONCRETE.

Getting Ready to Build

Before a dam is built, scientists take samples, or small parts, of the ground. These samples help them understand if the ground will be strong enough to hold the dam's weight. The taller the dam, the stronger the ground underneath it needs to be.

Engineers and scientists work together to make models of dams. The model can show them how the dam will **react** in certain conditions, such as **earthquakes**. A model is often created using a computer program or certain types of matter, like wood or rubber.

Building Blocks

Dam models help engineers understand what problems may come up and how to fix them before a dam is built.

THIS TESTING OF A MODEL DAM TOOK PLACE AT COLORADO STATE UNIVERSITY IN 1957.

Building the Dam

One of the first steps when building a dam is to change where the river's water flows. The riverbed, or ground at the river's bottom, needs to be dry to build a dam. To do this, a tunnel is built around the dam's construction site. The tunnel **diverts** the river away.

The next step is to remove loose rocks from the riverbed. The dam may fall down otherwise. Then, the strong plinth, or concrete base of the dam, is built into the solid ground.

Building Blocks

Engineers may build something similar to a small dam to create a dry area around a dam's construction site. This circular or rectangular structure is called a cofferdam.

TO BUILD A DAM, CONSTRUCTION WORKERS HAVE TO USE BIG MACHINES, SUCH AS CRANES, DIGGERS, AND DUMP TRUCKS.

Steel and Concrete

Now, the main structure of the dam can be built. This tall structure needs a lot of concrete and steel to make it strong enough to hold back water. A dam is usually built with a surface that curves in, which helps the dam take in the steady pressure of water it must withstand.

If a dam also acts as a hydroelectric plant, engineers may have to build structures to fit that purpose. This often requires even more planning and materials.

Building Blocks

Some concrete dams also have steel bars inside their concrete. This makes the dams even stronger.

IT CAN TAKE MANY YEARS TO BUILD A DAM. THE LONGER IT TAKES TO BUILD A DAM, THE MORE MONEY IT OFTEN COSTS.

Hydroelectric Power

Hydroelectric power is electricity made using the **energy** of flowing water. When water flows through a dam, it falls due to gravity, or the force that pulls objects toward Earth's center. In some dams, this falling water then turns a turbine, or a fanlike machine. When the turbine spins, it creates electricity we can then use to power our homes.

Hydroelectricity is one of the main types of electricity we use. It's one of the cleanest sources of electricity because it doesn't create waste and is **renewable**.

Building Blocks

In 2018, hydroelectric power made up around 16 percent of all electricity produced worldwide!

IN THE UNITED STATES, THE HOOVER DAM ON THE ARIZONA-NEVADA BORDER HELPS CREATE HYDROELECTRIC POWER.

Famous Dams

The Three Gorges Dam on the Yangtze River in China is a gravity dam and one of the world's most useful hydroelectric dams with 32 turbines. Construction on the dam began in 1994 and was completed in 2006.

However, many dams get taken down. Dams are usually taken down so plants and animals in the area can live and grow. For example, two dams on Washington's Elwha River were removed starting in 2011. This allowed the river to flow freely again so the surrounding natural area could grow.

Building Blocks

The building of a dam can make it harder—or impossible—for fish, such as salmon, to get to the place where they have their babies.

SADLY, THE THREE GORGES DAM DISPLACED MORE THAN 1.3 MILLION PEOPLE. THIS MEANS THE BUILDING OF THE DAM FLOODED AREAS WHERE MANY PEOPLE LIVED.

Make Your Own Dam

Now that you know how dams are built, let's use sand, small rocks, and wooden sticks to make your own dam!

What You Need:

- long, shallow, clear food storage box
- sand
- small rocks (such as rocks for an aquarium)
- wooden sticks
- bucket of water

How to:

1. Fill the food storage box with sand.
2. Dig out a long, narrow area that cuts through the sand from one side of the container to the other. This creates your riverbed.
3. Pick a site for your dam in the storage box. Then, place wooden sticks into the sand to make a wall. Place more wooden sticks about half an inch (1.3 cm) farther down the "river." Fill in the gap between the two sets of sticks with a mixture of wet sand and small rocks. This is your dam!
4. Finally, pour water into your riverbed from the top end of the storage box. Does your dam stop the flow of water?

Glossary

divert: to alter or change the direction of something

earthquake: a shaking of the ground caused by the movement of Earth's crust

energy: power used to do work

engineer: someone who uses science and math to build better objects

environment: the natural world in which a plant or animal lives

hydroelectric: having to do with creating power by using the movement of water

irrigation: the watering of a dry area by man-made means in order to grow plants

react: to change in a certain way when something happens

renewable: able to be replaced by nature

reservoir: a lake that is usually man-made and supplies water for communities of people

For More Information

Books

Krajnik, Elizabeth. *Making Dams and Reservoirs.* New York, NY: PowerKids Press, 2019.

Pettiford, Rebecca. *Dams.* Minneapolis, MN: Jump! Inc., 2016.

Yasuda, Anita. *Canals and Dams! With 25 Science Projects for Kids.* White River Junction, VT: Nomad Press, 2018.

Websites

Building Big: Dams
www.pbs.org/wgbh/buildingbig/dam
Read more about building dams and famous dams around the world on this website.

Engineering Facts
www.sciencekids.co.nz/sciencefacts/engineering/dams.html
Check out some more interesting facts about dams.

Types of Dams
www.ussdams.org/dam-levee-education/overview/types-of-dams
Learn about the different types of dams.

Publisher's note to educators and parents: Our editors have carefully reviewed these websites to ensure that they are suitable for students. Many websites change frequently, however, and we cannot guarantee that a site's future contents will continue to meet our high standards of quality and educational value. Be advised that students should be closely supervised whenever they access the internet.

Index